Practical Guide To Peptides For Pharmacists

Dr. Lisa Faast

Copyright © 2025 Dr. Lisa Faast | DiversifyRx
All rights reserved. No part of this publication may be reproduced, distributed, or transmitted in any form or by any means, including photocopying, recording, or other electronic or mechanical methods, without the prior written permission of the author, except in the case of brief quotations embodied in critical reviews and certain other noncommercial uses permitted by copyright law.

This book is independently published.
drlisafaast.com
info@diversifyrx.com
ISBN: 979-8-9925160-1-2
First Edition | February 11, 2025 | USA

Disclaimer: This book is for educational and informational purposes only. The content provided does not constitute medical advice, nor does it replace consultation with a qualified healthcare professional. Some peptides discussed in this book may not be FDA-approved or legally available in all jurisdictions. Readers are encouraged to conduct their own research and consult with a licensed medical professional before using any substances mentioned herein. The author and publisher assume no responsibility for the accuracy, effectiveness, or use of the information provided. Legal status and approvals can change frequently. Always refer to the most recent guidelines from the FDA or your state board of pharmacy.

For more information about this book, visit peptidebook.info

Table of Contents

Introduction 1
   Peptides are changing the world 1
A Look Around The Corner 8
   The Peptides Are Coming! The Peptides Are Coming! 8
      FDA-Approved Peptide Drugs
      Peptide Drugs in Development
      Recent FDA Approvals
What Exactly Is A Peptide?. 4
Size Does Matter 6
      Exactly How Different Is A Peptide From A Protein?
What Information Is Ahead? 8
Quick Reference Chart 10
Popular Peptides By Indication 11
Peptide Details 12
   What You Need & Want To Know
   What's Included
   Criteria For Being Able To Be Compounded
AICAR
   Acadesine 14
AOD 9064
   Hexadecapeptide | Awesome Obesity Drug 15
BPC-157
   Pentadecapeptide, Body Protection Compound. 16
Bremelanotide
   PT-141, Vyleesi 17
Cagrilintide
   Long Acting Amylin Analog. 18
CJC-1295
   Growth Hormone-Releasing Factor. 19
Dihexa
   N-hexanoic-Tyr-Ile-(6) aminohexanoic amide. 20
DSIP
   Delta Sleep-Inducing Peptide 21

Table of Contents

Follistatin 344
FS 22

GHK-Cu
Copper Peptide 23

GHRP-2
Pralmorelin | Growth Hormone Releasing Peptide-2 24

GHRP-6
Growth Hormone-Releasing Hexapeptide 25

Hexarelin
Examorelin 26

Ibutamoren
(MK-677) 27

Ipamorelin
Growth Hormone Secretagogue 28

Liraglutide
GLP-1 29

Melanotan II
MT2 30

MOTS-c
Mitochondrial Derived Peptide 31

Oxytocin
Pitocin, Love Hormone 32

Petrelintide
ZP8396 33

Retatrutide
Triple GLP-1, GIP, Glucagon Agonist 34

Selank
TP-7 35

Semaglutide
Ozempic & Wegovy 36

Semax
Semaxum 37

Sermorelin
Geref 38

Table of Contents

TB-500
    THYMOSIN Beta-4 39

Tesamorelin
    Egrifta 40

Tesofensine
    Serotonin–Noradrenaline–Dopamine Reuptake Inhibitor 41

Tirzepatide
    GLP-1 and GIP 42

Zinc Thymulin
    Serum Thymic Factor 43

Popular Peptide Stacks 44

OTC, API, 503Bs
    Additional Resources To Support Your Patients 47

Bonus!
    Comprehensive Alternative Medicine Options Chart 48
    Additional Alternative Medicine Options Chart 50

About The Author 54

Introduction

Peptides are changing the world.

Welcome to the *Practical Guide to Peptides for Pharmacists!*

If you're holding this book, it means you're either a forward-thinking pharmacy badass ready to take your practice to the next level, or you're simply curious about one of the most exciting frontiers in health and wellness today. Either way, you've come to the right place.

Howdy! I'm Dr. Lisa Faast, a pharmacist who has spent my career fighting for the success of independent pharmacies and empowering pharmacists like you to expand your impact. From consulting NFL teams to owning pharmacies to running DiversifyRx, my passion lies in helping others unlock their full potential in the healthcare space.

Peptides have fascinated me for years, and not just because they're trendy or buzzy, but because they work. And trust me, as a pharmacist, you're uniquely positioned to harness their power to transform the health of your patients—and your business.

So, why peptides?

These tiny chains of amino acids might not look like much on paper, but they're rewriting the rules of healthcare. Peptides are already being used to optimize athletic performance, improve metabolic health, combat aging, and even tackle complex diseases. Their potential is enormous, and we're only scratching the surface of what they can do. This isn't just a passing fad—this is the future of personalized medicine.

But let's be real: navigating the peptide world can feel like stepping into uncharted territory. You've got questions, and this book has answers. From understanding how peptides work to integrating them *(legally)* into your pharmacy practice or to just being able to answer questions you get from your patients or family, this guide is here to demystify the science, provide actionable steps, and set you up for success. Whether you're looking to help your patients boost their energy, recover faster, or manage chronic conditions more effectively, peptides offer a powerful tool to expand your services and your expertise.

> *We pharmacists are often the unsung heroes of healthcare, bridging the gap between science and patient care. Peptides allow us to take that role even further, offering cutting-edge solutions that genuinely change lives.*

Ready to dive in?

Let's explore the practical, powerful, and promising world of peptides together. This isn't just about keeping up with trends; it's about staying ahead, leading the charge, and showing the world what pharmacists can really do.

Disclaimer: This book is for informational purposes only and is not intended to provide medical advice, diagnosis, or treatment. The information provided may not reflect the most current research or regulations, as details about peptides and their uses can change frequently. Always consult with a qualified healthcare professional before starting any new treatment or supplement regimen. We encourage you to do your own research and verify the information presented here to ensure it aligns with your individual needs and circumstances. Continuing to use this document means you agree to this statement. The legal landscape of pharmaceuticals can change rapidly. This information is current as of the publication date. Always refer to the latest information available from the FDA or your state board of pharmacy.

A Look Around The Corner

The Peptides Are Coming! The Peptides Are Coming!

As of the beginning of 2025, the landscape of peptide-based therapeutics is rapidly expanding, reflecting significant advancements in pharmaceutical research and development.

FDA-Approved Peptide Drugs
- **Current Approvals:** Approximately 60+ peptide drugs have received FDA approval, addressing a diverse range of medical conditions.

Peptide Drugs in Development
- **Clinical Trials:** Around 140 peptide-based drugs are currently undergoing clinical trials, indicating a robust pipeline of potential future therapies.
- **Preclinical Development:** Over 500 peptide drugs are in preclinical stages, underscoring the growing interest and investment in peptide therapeutics.

Recent FDA Approvals
- **2023:** The FDA approved nine TIDES (therapeutic peptides and oligonucleotides), representing 16% of the year's new drug approvals.
- **2022:** Five TIDES received FDA approval, including four peptides and one oligonucleotide, accounting for 14% of new drug approvals.
- **2021 Approvals:** In 2021, the FDA approved ten TIDES, comprising eight peptides and two oligonucleotides, reflecting the growing significance of this class in therapeutic development.
- **2020 Approvals:** Despite the challenges posed by the COVID-19 pandemic, the FDA approved six TIDES in 2020, maintaining the momentum in peptide and oligonucleotide drug development.

 ***Insight:*** *One of the most anticipated peptide approvals coming in 2026 is Retatrutide from Eli Lilly and Company. I think the weight loss craze has only just begun!*

Sources:
- Peptides Guide, FDA-Approved Peptide Drugs and Clinical Pipeline, peptidesguide.com
- MDPI, Recent Progress of Peptides in FDA Approvals: TIDES in 2022 and 2023, mdpi.com
- MDPI, Peptides and Oligonucleotides in FDA Approvals, mdpi.com
- PLOS ONE, Overview of FDA Approvals of Peptides and Therapeutic Proteins (1980s-2009). Available at: journals.plos.org

For the most current information, consulting the FDA's official resources is recommended.

But...
What Exactly Is A Peptide?

Before we dive into how peptides can revolutionize your pharmacy practice, let's answer the burning question: what is a peptide, anyway?

Simply put, peptides are short chains of amino acids, the building blocks of proteins. Think of them as mini-proteins, but with a much more targeted and specific function. While proteins are like a symphony orchestra, peptides are the soloist—focused, precise, and designed to perform a specific task.

Peptides occur naturally in the body and play vital roles in almost every biological process. From regulating hormones to healing wounds and signaling the immune system, these little powerhouses are integral to keeping us alive and well. Scientists have learned to harness and replicate peptides to create therapies that mimic or enhance these natural processes, opening up a world of therapeutic possibilities.

> *From regulating hormones to healing wounds and signaling the immune system, these little powerhouses are integral to keeping us alive and well.*

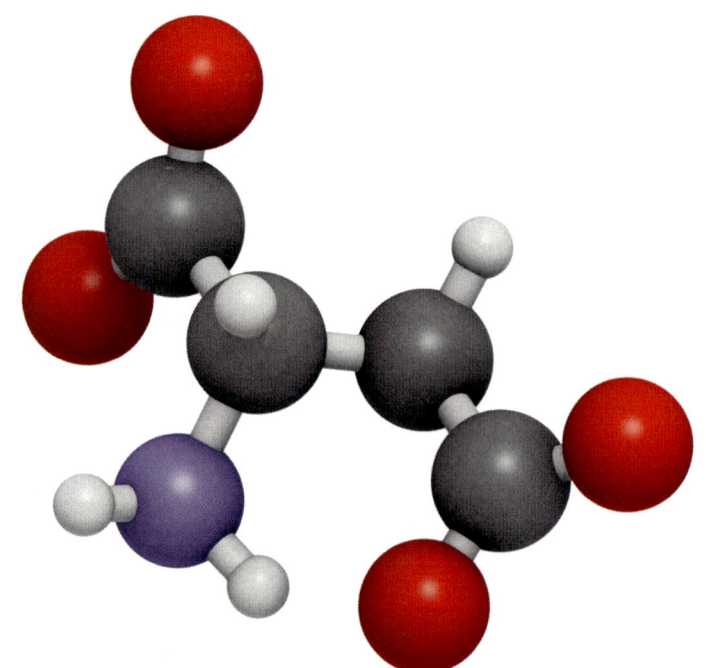

The beauty of peptides lies in their specificity. Unlike traditional medications that can have broad and sometimes unwanted effects, peptides are designed to interact with specific receptors in the body.

This targeted approach means they often come with fewer side effects and greater effectiveness—a win-win for both patients and practitioners.

Here's an example: insulin, one of the most well-known peptides, has been saving lives for nearly a century. Today, researchers are developing peptides to tackle everything from obesity and chronic pain to cancer and neurodegenerative diseases. The potential applications are staggering, and as pharmacists, we're at the forefront of making these advancements accessible to our patients.

> *The beauty of peptides lies in their specificity. Unlike traditional medications that can have broad and sometimes unwanted effects, peptides are designed to interact with specific receptors in the body.*

In short, peptides are the unsung heroes of modern medicine. They're precise, powerful, and poised to reshape the future of healthcare. And now, it's your turn to learn how to incorporate them into your practice.

Many pharmacists, including myself, would love to be able to dispense or compound peptides pursuant to a prescription. Even if you are unable to dispense, you can sell many over-the-counter or even refer your patients to other legitimate sources.

Size Does Matter

Exactly How Different Is A Peptide From A Protein?

Peptides vs. Proteins: Size and Weight Comparison

To truly appreciate peptides and their applications, it's helpful to understand how they differ from proteins, particularly in size and weight.

- **Size:** Peptides are short chains of amino acids, typically consisting of 2 to 50 amino acids. Each amino acid is about 0.36 nanometers (nm) long, so a peptide might range from ~0.7 nm to 18 nm in length. In contrast, proteins are much larger, often composed of hundreds to thousands of amino acids. For example, hemoglobin (a protein with 574 amino acids) is approximately 5.5 nm in diameter, while titin, a muscle protein with over 30,000 amino acids, can stretch up to 1 micron in length.

- **Weight:** Peptides have molecular weights typically between 200 and 5,000 Daltons (Da), depending on their length and composition. Proteins, on the other hand, are significantly heavier. Hemoglobin weighs around 64,500 Da, and titin, one of the largest human proteins, exceeds 3 million Da.

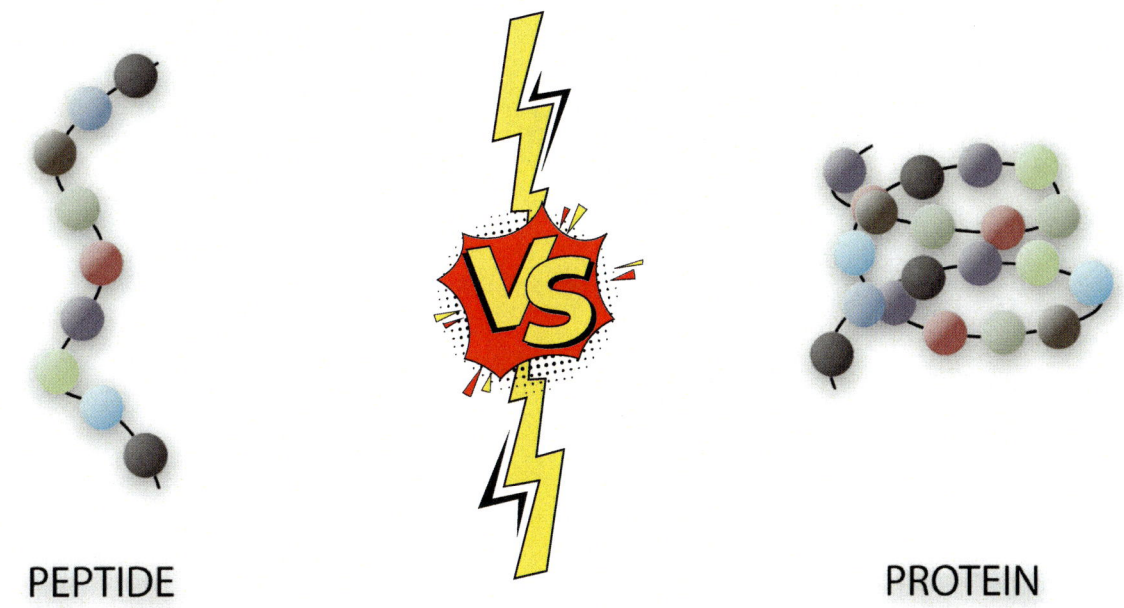

PEPTIDE   PROTEIN

Proteins often fold into intricate three-dimensional structures, while peptides are simpler, making them lighter and smaller.

Comparison: A peptide like insulin (a hormone and peptide) weighs about 5,800 Da and is composed of 51 amino acids, whereas a protein like hemoglobin is over 11 times larger in molecular weight and significantly more complex in structure. Proteins often fold into intricate three-dimensional structures, while peptides are simpler, making them lighter and smaller.

This stark contrast in size and weight highlights why peptides are often used for more targeted and specific roles in the body, while proteins handle diverse and complex biological functions. Understanding these differences helps us appreciate the versatility and potential of peptides in modern medicine.

What Information Is Ahead?

I called this book "Practical" for a reason.

A comprehensive textbook on all possible clinically relevant peptides would be the size of an 80's Encyclopedia Brittanica.

You would have tons of information, and it would not be practically useful for your day-to-day pharmacist life.

I whittled a massive list of hundreds down to the most practical ones. The filtering process was based on those you are most likely to encounter as a typical independent or retail pharmacist. These are the ones that you will get questions about from patients and providers alike.

Left out were peptides that play in the world of cancer, immunology, and rare diseases. Included were those in the world of building muscle, losing fat, and looking better.

While the FDA doesn't regulate the practice of medicine, it most certainly regulates the pharmaceutical world, including peptides. The FDA has NOT approved most clinically relevant peptides. It's anybody's guess what might happen with the new administration coming in during 2025.

Then there are the "research" peptides. Most pharmacists I know get really angry about this class, which I can sympathize with. A majority of the peptides that your patients will ask you about will come from this realm. While you may not like it, you should be aware of it and what is offered.

Research chemicals are NOT approved for use in humans. This class exists as chemicals need to be studied in the animal realm long before they are made available to humans. These companies can be high-quality or bottom-of-the-barrel companies.

> *Patients will often ask for your opinion. Rather than leave them to their own devices, when you are armed with the right knowledge, you can help them make better choices even if you can't be the one to provide the treatment.*

Also, don't forget that we live in a very globalized world. Your patients can easily get lots of treatment options that are not available in the US. The more informed you are, the better you will be able to help and guide them.

The information in the following pages should equip you with the right foundational knowledge so you can intelligently answer questions, know if you can legally compound an ingredient, and be prepared to help your patients.

Quick Reference Chart

Peptide Name	Dosage Form	Dose Range	Uses	Daltons Weight	# of Amino Acids	Ever FDA Approved?
AICAR	INJ	Variable	AMPK activation	338.221	n/a	Not FDA-approved.
AOD9604	INJ	0.25-1 mg daily	Fat metabolism	1815.1	15	Not FDA-approved.
BPC-157	INJ	200-400 mcg daily	Tissue repair	1419.535	15	Not FDA-approved.
Bremelanotide	INJ	1.75 mg as needed	Libido enhancement	1025.2	7	Yes, FDA-approved in 2019 as **Vyleesi** for HSDD.
Cagrilintide	INJ	0.3-2.4 mg weekly	Obesity	3422.5	31	Not FDA-approved (under clinical trials).
CJC-1295	INJ	100-200 mcg daily	GH secretion	3647.29	30	Not FDA-approved.
Dihexa	Capsule	1-10 mg daily	Cognitive function	615.8	3	Not FDA-approved.
Follistatin 344	INJ	Variable	Muscle growth	39000	344	Not FDA-approved.
GHK-Cu	Cream	Variable	Wound healing	340.8	3	Not FDA-approved.
Ipamorelin	INJ	100-300 mcg daily	GH secretion	711.85	5	Not FDA-approved.
Liraglutide	INJ	0.6-3 mg daily	Weight management	3751.2	37	Yes, 2010 as **Victoza** for type 2 diabetes, 2014 as **Saxenda** for weight management.
Melanotan II	INJ	0.25-1 mg daily	Skin pigmentation, libido enhancement	1024.2	7	Not FDA-approved.
Oxytocin	INJ/Nasal	Variable	Labor induction, social bonding	1007.2	9	Yes, FDA-approved in 1954 as **Pitocin** for labor induction.
Petrelintide	INJ	Variable	Appetite suppression	1912.1	16	Not FDA-approved.
Retatrutide	INJ	2 mg - 12 mg weekly	Weight management, blood glucose regulation	1679.29	39	Not FDA-approved (under clinical trials).
Selank	Nasal Spray	Variable	Anti-anxiety, cognitive enhancement	828.9	7	Not FDA-approved.
Semaglutide	INJ	0.25-2.4 mg weekly	Weight management, blood glucose regulation	4113.58	31	Yes, 2017 as **Ozempic** for type 2 diabetes, 2021 as **Wegovy** for weight management.
Semax	Nasal Spray	Variable	Neuroprotection, cognitive enhancement	828.9	7	Not FDA-approved.
Sermorelin Acetate	INJ	200-500 mcg daily	GH secretion	3357.88	29	Not FDA-approved.
TB-500	INJ	Variable	Tissue repair, inflammation	4963.4	43	Not FDA-approved.
Tesamorelin	INJ	2 mg daily	Reduce abdominal fat in HIV associated lipodystrophy	5135.8	44	Yes, FDA-approved in 2010 as **Egrifta** for HIV-associated lipodystrophy.
Tesofensine	Capsule	0.25-0.5 mg daily	Reduce appetite, weight loss	308.43	36	Not FDA-approved.
Tirzepatide	INJ	2.5-15 mg weekly	Weight loss, blood glucose regulation	4813.5	39	Yes, FDA-approved in 2022 as **Mounjaro** for type 2 diabetes.
Zinc Thymulin	INJ	Variable	Hair growth, immune modulation	Variable	9	Not FDA-approved.

Popular Peptides By Indication

Peptides for Hair Loss
- **Zinc Thymulin:** Promotes hair growth and has immune-modulating properties.
- **BPC-157:** May support scalp tissue repair and reduce inflammation.
- **GHK-Cu:** Stimulates hair growth through collagen production and wound healing.
- **TB-500 Thymosin Beta-4:** Supports tissue repair and could improve scalp health.

Peptides for Weight Loss
- **AOD9604:** Stimulates fat metabolism and reduces fat storage.
- **Semaglutide:** A GLP-1 agonist that aids in weight management & blood sugar.
- **Tirzepatide:** Targets multiple pathways for weight loss & glucose control.
- **Tesofensine:** Suppresses appetite and enhances metabolic health.
- **Retatrutide:** Combines appetite regulation & metabolic improvements.
- **Cagrilintide:** Works alongside other weight loss peptides to amplify results.

Peptides for Building Muscle
- **CJC-1295 + Ipamorelin:** Boosts GH release for muscle growth and recovery.
- **GHRP-6:** Stimulates appetite and growth hormone secretion for muscle gains.
- **Hexarelin:** Potent growth hormone secretagogue for muscle repair.
- **Tesamorelin:** Enhances muscle growth while targeting abdominal fat.
- **IGF-1 LR3:** Directly promotes muscle hypertrophy and tissue regeneration.
- **Follistatin 344:** Inhibits myostatin, promoting muscle growth.

Peptides for Wound Healing & Tissue Repair
- **BPC-157:** Accelerates healing of muscles, tendons, and ligaments.
- **TB-500 (Thymosin Beta-4):** Enhances tissue repair and reduces inflammation.
- **GHK-Cu:** Supports skin regeneration and wound healing.
- **Thymosin Alpha-1:** Combines immune modulation with healing properties.

Peptides for Cognitive Enhancement
- **Selank:** Reduces anxiety and improves cognitive focus.
- **Semax:** Enhances neuroprotection and supports memory and focus.
- **Dihexa:** Potent cognitive enhancer for improving brain function.
- **CJC-1295 + Ipamorelin:** Indirectly supports cognitive health through improved sleep and recovery.

Peptides for Skin and Anti-Aging
- **Melanotan II:** Enhances skin pigmentation and protects against UV damage.
- **GHK-Cu:** Collagen synthesis & skin healing.
- **PT-141:** Combines libido & skin benefits.
- **AOD9604:** Supports anti-aging effects through improved metabolism.

Peptides for Libido & Reproductive Health
- **Kisspeptin-10:** Enhances reproductive hormone regulation.
- **PT-141:** Directly targets arousal and libido.
- **Melanotan II:** Improves libido & skin.
- **Oxytocin:** Enhances bonding & intimacy.

Peptide Details

What You Need & Want To Know

There's a ton of information out there...

And not all of it is useful to you as a pharmacist trying to help patients, speak with providers, or discuss with colleagues.

That's why I have gathered the most important *AND PRACTICAL* information on each peptide for you. It is what you need to know.

As always, please do your own research for unique circumstances or for more detailed information.

I have arranged this section in alphabetical order to make it easy to find what you are looking for.

For the detailed peptide sections, I have included only the "popular commercialized" peptides. This way, you are not thumbing through a bunch of pages you will never need.

Here is the list of peptides that have a detailed page:

AICAR	GHK-Cu	Retatrutide
AOD 9604 (hexadecapeptide)	GHRP-2	Selank
	GHRP-6	Semaglutide
BPC-157	Hexarelin	Semax
Bremelanotide (PT-141)	Ibutamoren (MK-677)	Sermorelin Acetate
Cagrilintide	Ipamorelin	TB-500
CJC-1295	Liraglutide	Tesamorelin
Dihexa	Melanotan II	Tesofensine
DSIP	MOTS-c	Tirzepatide
Follistatin 344	Petrelintide	Zinc Thymulin

Peptide Details

What's Included

Information Included For Each Peptide:

- Common name
- Other name
- Benefits
- General description
- Common dosing or protocols
- Potential side effects
- Can it be compounded by pharmacies
- References

Criteria For Being Able To Be Compounded

This information may change at any time. Be sure you understand your state and FDA rules. This is not an all-inclusive criteria to be compounded but rather a simple set of rules to follow so you can easily determine if you can make a compound with a certain API (active pharmaceutical ingredient).

- The ingredient has been a part of an FDA-approved drug at some point and was not withdrawn over safety concerns.
- The ingredient is on the FDA's "ok to compound" approved list.
- The ingredient is NOT on the FDA's "do not compound" list.
- The ingredient has a USP/NF monograph.

Refer to your state board of pharmacy and FDA regulations.
https://www.fda.gov/media/174456/download
https://www.fda.gov/media/94402/download

Don't forget - if a product is under patent, chances are you cannot compound it from an API, even in a different dosage form. Patent laws are very strong. Products listed as "Currently In Shortage" on the FDA shortage website can be compounded even if they are under patent.
https://dps.fda.gov/drugshortages

AICAR
Acadesine

Benefits May Include:
- Activation of AMPK
- Increased metabolic activity
- Enhanced fat oxidation
- Improved insulin sensitivity
- Increase athletic performance (Banned in sports)

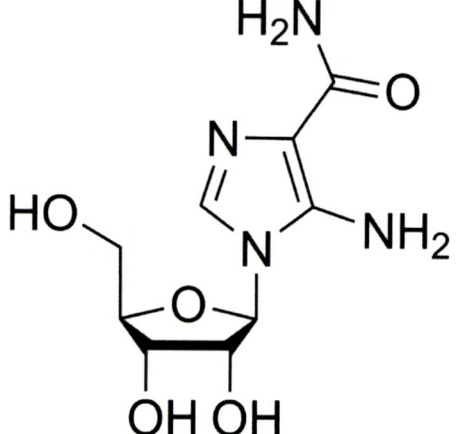

AICAR is a nucleotide analog and AMPK activator. It was initially developed for metabolic and cardiovascular conditions but has since been researched for its potential in athletic performance and fat metabolism. AICAR works by mimicking conditions of energy deprivation, thereby activating AMPK, which plays a key role in regulating energy balance. Studies suggest it may aid in weight management, fat metabolism, and improving insulin sensitivity.

Common Dosages & Protocols:
- Injection: Daily SQ dose of 25mg for 2 weeks up to twice a year with at least 2 months in between cycles.
- Referred to as 'exercise in a bottle'.

Potential Side Effects:
- High doses have led to kidney dysfunction
- Lactic acidosis
- Hypoglycemia
- Fatty liver disease
- Irritable bowel syndrome
- Disrupted hemodynamics

Clinical References:
- Canto C, Auwerx J. "AMPK activators as a therapeutic strategy for metabolic disorders." Nat Rev Mol Cell Biol. 2010.
- Narkar VA, Downes M, Yu RT, et al. "AMPK and PPARdelta agonists are exercise mimetics." Cell. 2008.
- https://www.peptides.org/aicar-dosage-calculator/

AOD 9064

Hexadecapeptide | Awesome Obesity Drug

Benefits May Include:
- Promotes fat metabolism without affecting blood glucose
- Supports cartilage and bone repair
- Reduces fat accumulation

AOD 9604 is a modified peptide fragment of human growth hormone (HGH) that specifically targets fat metabolism. Unlike HGH, it does not promote muscle growth or increase IGF-1 levels, making it ideal for weight loss without systemic effects. It is being researched for obesity treatment, cartilage repair, and anti-inflammatory properties. Preclinical and clinical studies support its role in fat oxidation and tissue repair.

Common Dosages & Protocols:
- Injection: 0.25-1 mg daily
- Cream: 600 mcg/g applied daily for targeted anti-inflammatory effects
- Oral 9mg, 27mg, 54mg (below 30mg to reduce GI distress)

Potential Side Effects:
- Mild to moderate headache
- Mild to moderate euphoria
- GI distress/diarrhea (oral dosing)

Clinical References:
- Heffernan M, Stier H. "Lipolytic effects of AOD9604 in obese subjects." Endocrinology. 2013.
- Ng FM. "Role of HGH analogs in metabolic disorders." Metabolism. 2006.
- https://jofem.org/index.php/jofem/article/view/157/194

BPC-157
Pentadecapeptide, Body Protection Compound

Benefits May Include:
- Accelerated wound healing
- Tissue regeneration
- Reduced inflammation
- Gastrointestinal protection

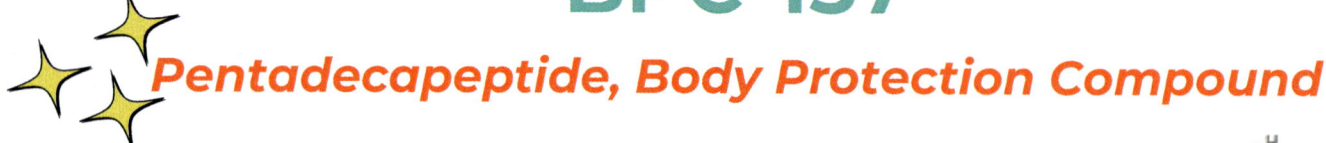

BPC-157 (Body Protection Compound-157) is a synthetic peptide derived from a protein found in the human stomach. It has been extensively studied for its regenerative properties, particularly in promoting wound healing and tissue repair. BPC-157 has shown promise in healing ligaments, tendons, muscles, and even nerve tissues. Additionally, it has protective effects on the gastrointestinal tract, reducing symptoms associated with inflammatory bowel disease (IBD) and ulcers.

Common Dosages & Protocols:
- Injection: 200-400 mcg daily, subcutaneous or intramuscular.
- Oral: Capsules of 500 mcg taken daily for gastrointestinal issues.

Potential Side Effects:
- Mild headaches
- Nausea
- Localized irritation at the injection site

Clinical References:
- Sikiric P, Seiwerth S. "The role of BPC-157 in wound healing and tissue repair." Curr Pharm Des. 2017.
- Wu X, Jing L. "BPC-157 and gastrointestinal protection: A comprehensive review." World J Gastroenterol. 2015.

Bremelanotide
PT-141, Vyleesi

Benefits May Include:
- Enhanced sexual arousal
- Improved libido
- Improved erections

Bremelanotide, also known as PT-141, is a melanocortin receptor agonist used to treat hypoactive sexual desire disorder (HSDD) in premenopausal women. FDA-approved under the brand name Vyleesi, it works by activating brain pathways associated with sexual desire.

Common Dosages & Protocols:
- Injection: 1.75 mg to 2 mg subcutaneously as needed.
- Nasal Spray: 1.75 mg to 2 mg 30-60 minutes before activity
- Used 1-2 times per week

Potential Side Effects:
- Nausea
- Flushing
- Headaches

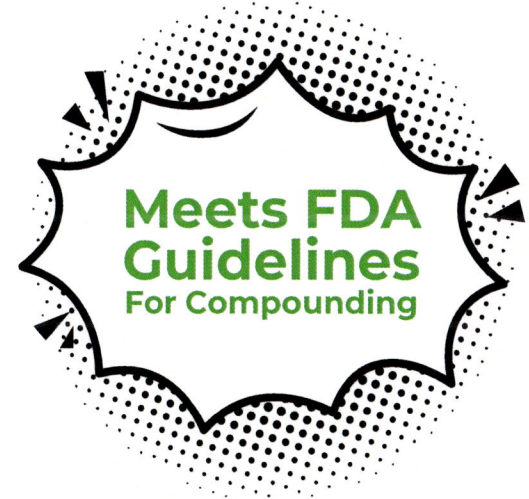

Clinical References:
- Clayton AH, Kingsberg SA. "Efficacy of Bremelanotide in HSDD." Obstet Gynecol. 2019.
- Safarinejad MR. "Melanocortin receptor agonists in sexual function." J Urol. 2008.

Cagrilintide
Long Acting Amylin Analog

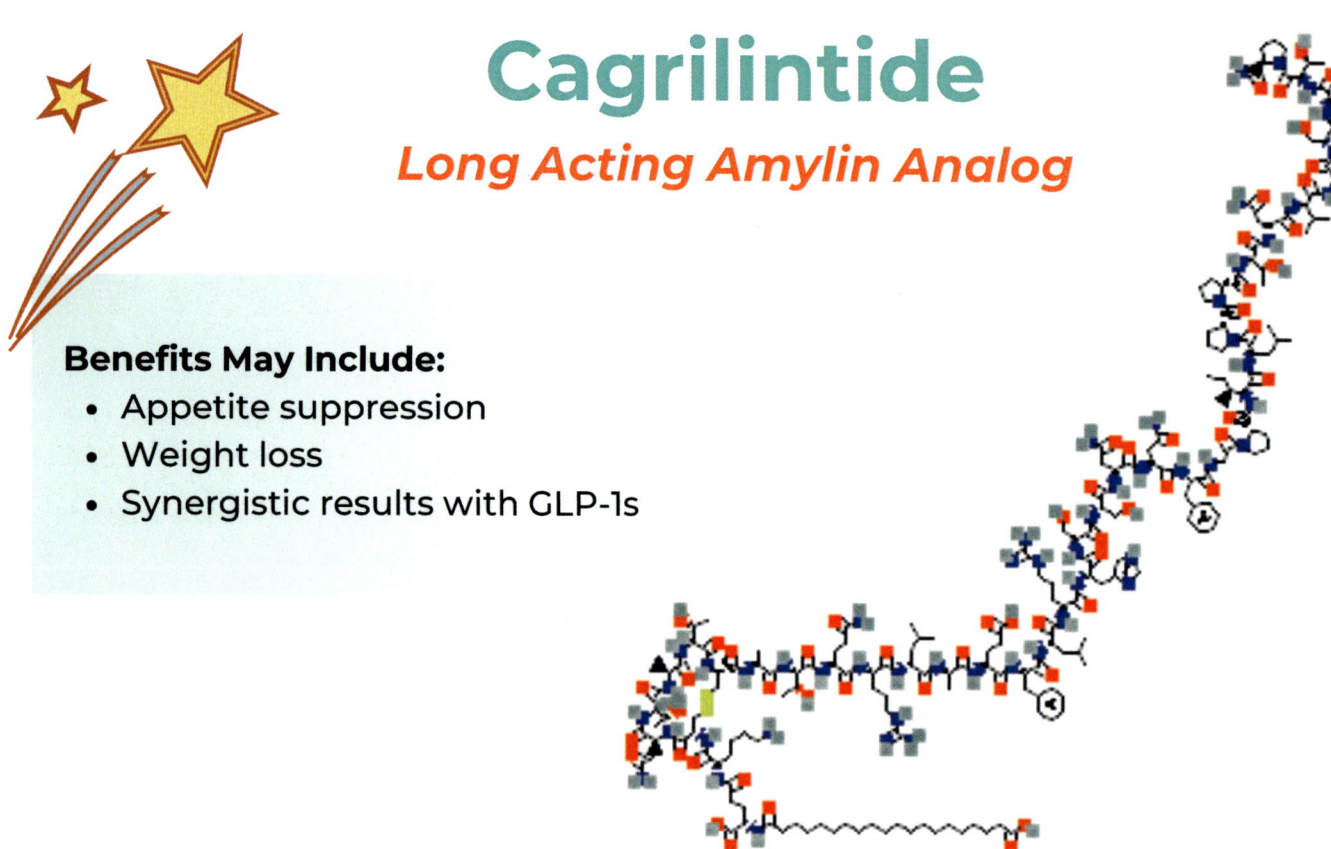

Benefits May Include:
- Appetite suppression
- Weight loss
- Synergistic results with GLP-1s

Cagrilintide is an investigational peptide analog of amylin that enhances feelings of fullness and reduces appetite. It is being studied in combination with GLP-1 receptor agonists for its synergistic effects in weight management. Novo's CagriSema is in Phase III and is expected to apply for approval towards the end of 2025.

Common Dosages & Protocols:
- Injection: Weekly dosing ranging from 0.3 to 2.4 mg.
- Can be used with GLP-1s like semaglutide

Potential Side Effects:
- Nausea
- Vomiting
- Diarrhea

Clinical References:
- Lau J, Bloch P. "Cagrilintide in obesity management." Lancet. 2021.
- Frias JP, Nauck MA. "Combination therapies for obesity." Obesity Research. 2020.

CJC-1295
Growth Hormone-Releasing Factor

Benefits May Include:
- Increased growth hormone secretion
- Enhanced muscle recovery
- Increased IGF-1 levels

H—Tyr-D-Ala-Asp-Ala-Ile-Phe-Thr-Gln-Ser-Tyr-Arg—Lys-Val-Leu-Ala-Gln-Leu-Ser-Ala-Arg-Lys-Leu-Leu-Gln-Asp-Ile-Leu-Ser-Arg

CJC-1295 is a synthetic analog of growth hormone-releasing hormone (GHRH) that stimulates the release of growth hormone (GH). Its extended half-life allows for fewer injections compared to other GH secretagogues.

Common Dosages & Protocols:
- Injection: 100-200 mcg subcutaneously 1-3 times per week.
- Often combined with Sermorelin, Ipamorelin

Potential Side Effects:
- Water retention
- Injection site irritation

Clinical References:
- Teichman SL. "CJC-1295 and growth hormone secretion." Endocrinology. 2006.
- Ip C, Ho RJ. "GH analogs in therapeutic settings." Curr Pharm Biotechnol. 2010.

Dihexa

N-hexanoic-Tyr-Ile-(6) aminohexanoic amide

Benefits May Include:
- Improved cognitive function
- Potential neuroprotective effects

Dihexa is a synthetic peptide designed to enhance learning and memory. It is believed to promote synaptogenesis by binding to hepatocyte growth factor (HGF) receptors, which play a key role in neural repair and cognitive health. While promising in preclinical research, it is not FDA-approved.

Common Dosages & Protocols:
- Capsules: 1-10 mg orally per day.

Potential Side Effects:
- Bleeding
- Bruising

Clinical References:
- Larson J, Sherman MA. "Dihexa and cognitive enhancement: Preclinical results." J Neurosci. 2013.
- Harding SD, Armstrong JD. "The neurotherapeutic role of HGF mimetics." Neuropharmacology. 2018.

DSIP
Delta Sleep-Inducing Peptide

Benefits May Include:
- Improved sleep quality
- Stress reduction
- Potential pain relief

DSIP, or Delta Sleep-Inducing Peptide, is a naturally occurring neuropeptide that has been studied for its role in promoting restful sleep and reducing stress. It acts on the central nervous system to regulate sleep cycles, modulate stress response, and potentially enhance pain tolerance.

Common Dosages & Protocols:
- Injection: 100-300 mcg subcutaneously before bedtime.

Potential Side Effects:
- Mild drowsiness
- Temporary injection site irritation

Clinical References:
- Graf MV. "DSIP and its role in sleep regulation." Neuroscience. 2009.
- Kline CE. "Therapeutic potential of DSIP in stress disorders." Curr Pharm Des. 2015.

Follistatin 344
FS

Benefits May Include:
- Increased muscle mass
- Myostatin inhibition

Structural diagram of Follistatin 344 peptide with $IC_{50} = 4.0\ \mu M$, showing Val→Phg and Trp→Phe substitutions.

Follistatin 344 is a naturally occurring protein that inhibits myostatin, a regulatory protein that limits muscle growth. By suppressing myostatin, Follistatin 344 enhances muscle hypertrophy and recovery. It has been studied for potential use in muscle-wasting diseases.

Common Dosages & Protocols:
- Injection: 100mcg once a day for 10-30 days

Potential Side Effects:
- Injection site reactions
- Immune responses in rare cases

Does NOT Meet FDA Guidelines For Compounding

Clinical References:
- Lee SJ. "Follistatin as a therapeutic tool for muscle disorders." Muscle Nerve. 2010.
- Amthor H, Hoogaars W. "Myostatin inhibition for muscle growth." Trends Endocrinol Metab. 2012.

GHK-Cu
Copper Peptide

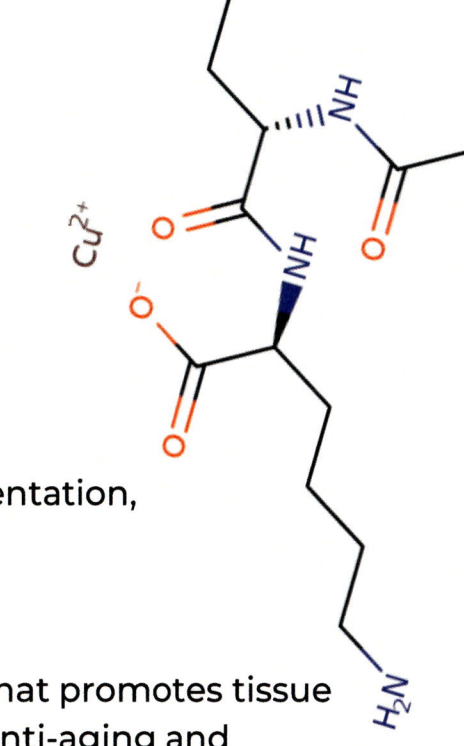

Benefits May Include:
- Accelerated wound healing
- Improved skin regeneration
- Increase hair growth and thickness
- Tighten loose skin
- Improve skin firmness, elasticity, and clarity
- Reduce fine lines and depth of wrinkles
- Smooth rough skin
- Reduce photodamage, mottled hyperpigmentation, skin spots and lesions

GHK-Cu is a naturally occurring copper peptide that promotes tissue repair and regeneration. It is commonly used in anti-aging and dermatological treatments.

GHK-Cu is a promising alternative treatment for hair loss. It may promote hair growth, strength, and volume, prevent thinning, and improve scalp health. It may also reduce lines and wrinkles, improve overall appearance, and increase skin density and thickness.

Common Dosages & Protocols:
- Topical creams: 0.05%-3% applied daily.
- Hair regrowth: up to 10%
- Oral dosages: 1 - 10mg daily.
- FDA Category 1 for oral and topical only

Potential Side Effects:
- Skin irritation

Clinical References:
- Pickart L. "GHK-Cu as a regenerative peptide." Med Hypotheses. 2009.
- Kang S, Chung JH. "GHK-Cu and wound healing." Dermatol Surg. 2015.

GHRP-2
Pralmorelin | Growth Hormone Releasing Peptide-2

Benefits May Include:
- Enhanced growth hormone secretion
- Improved muscle recovery
- Increased fat metabolism

GHRP-2 is a synthetic peptide that stimulates the release of growth hormone by acting on the ghrelin receptor in the hypothalamus. It is widely used in research for its ability to increase growth hormone levels, promoting muscle recovery, fat loss, and overall metabolic health.

Common Dosages & Protocols:
- Injection: 100-300 mcg subcutaneously 1-3 times daily.

Potential Side Effects:
- Increased appetite
- Temporary fatigue
- Mild water retention

Clinical References:
- Kaji H, Chihara K. "Mechanisms of growth hormone secretion by GHRPs." Endocrinology. 1997.
- Smith RG. "The therapeutic potential of GHRPs." Horm Res. 2001.

GHRP-6
Growth Hormone-Releasing Hexapeptide

Benefits May Include:
- Increased growth hormone secretion
- Enhanced muscle growth and fat loss

GHRP-6 is a synthetic peptide that binds to the ghrelin receptor, stimulating the release of growth hormone. It is often used in research and athletic recovery for its anabolic effects, including improved muscle growth, enhanced fat metabolism, and faster recovery from injuries.

Common Dosages & Protocols:
- Injection: 100-300 mcg subcutaneously 1-3 times daily.

Potential Side Effects:
- Increased appetite
- Temporary water retention
- Redness at the injection site

Clinical References:
- Garcia JM, Cata JP. "Growth hormone-releasing peptides in therapy." Expert Opin Investig Drugs. 2007.
- Smith RG. "The effects of GHRP-6 on anabolic processes." Horm Res. 2005.

Hexarelin
Examorelin

Benefits May Include:
- Increased growth hormone release
- Enhanced muscle recovery
- Improved fat metabolism

Hexarelin is a synthetic growth hormone secretagogue and ghrelin receptor agonist. It stimulates the pituitary gland to release growth hormone, promoting muscle growth, fat metabolism, and recovery. Hexarelin is widely studied in research settings for its potential therapeutic effects in growth hormone deficiency and cardiac health.

Common Dosages & Protocols:
- Injection: 100-200 mcg subcutaneously 1-2 times daily.

Potential Side Effects:
- Increased appetite
- Temporary water retention
- Mild injection site irritation

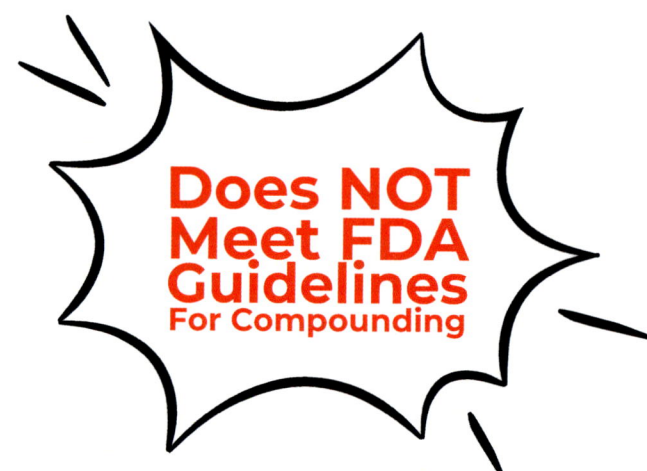

Clinical References:
- Smith RG. "Hexarelin and its effects on growth hormone secretion." J Endocrinol. 1997.
- Broglio F. "Hexarelin and metabolic health." Endocr Rev. 2009.

Ibutamoren
(MK-677)

Benefits May Include:
- Enhanced growth hormone secretion
- Increased muscle mass
- Improved bone density
- Improved sleep

Ibutamoren, also known as MK-677, is a non-peptide growth hormone secretagogue that mimics the action of ghrelin. It stimulates the release of growth hormone and insulin-like growth factor 1 (IGF-1) without impacting cortisol levels. This compound is commonly studied for its potential to treat conditions like muscle wasting, osteoporosis, and age-related hormonal decline.

Common Dosages & Protocols:
- Oral: 10-25 mg daily.

Potential Side Effects:
- Increased appetite
- Mild water retention
- Transient muscle pain

Clinical References:
- Nass R, Pezzoli SS. "Ibutamoren and growth hormone secretion." J Clin Endocrinol Metab. 2008.
- Smith RG. "The potential therapeutic applications of MK-677." Curr Opin Clin Nutr Metab Care. 2013.

Ipamorelin
Growth Hormone Secretagogue

Benefits May Include:
- Improved skin
- Increased muscle mass
- Improved recovery
- Better sleep
- Improved mental clarity
- Increased energy
- Stronger bones
- Improved metabolism

Ipamorelin is a selective growth hormone secretagogue that stimulates the release of growth hormone without significantly affecting other hormones like cortisol or prolactin. It is often used for its regenerative and anti-aging effects.

Common Dosages & Protocols:
- Injection: 100-300 mcg daily, subcutaneous.
- Often used with CJC-1295

Potential Side Effects:
- Injection site irritation
- Headaches
- Mild water retention

Clinical References:
- Raun K. "Ipamorelin: Growth hormone releasing properties." Curr Opin Endocrinol. 2005.
- Smith RG. "Clinical applications of growth hormone secretagogues." Curr Pharm Des. 2012.

Liraglutide
GLP-1

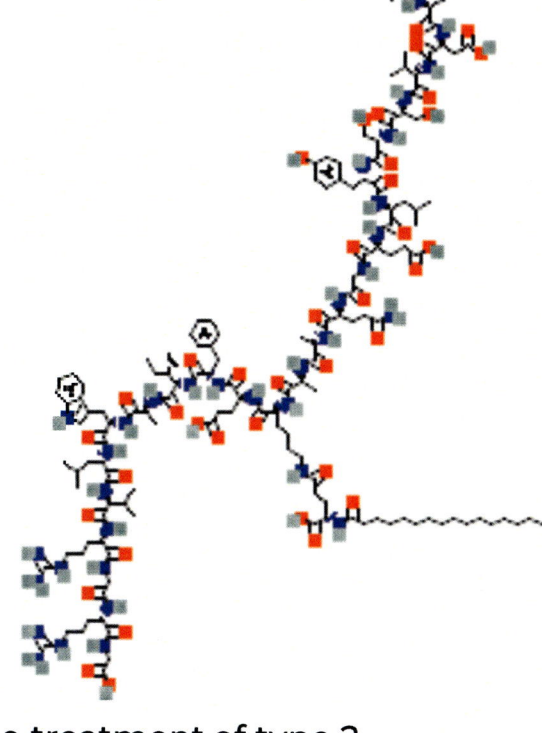

Benefits May Include:
- Improved blood glucose control
- Weight management
- Appetite suppression
- Reduce the risk of cardiovascular events

Liraglutide is a GLP-1 receptor agonist used in the treatment of type 2 diabetes and chronic weight management. It is FDA-approved under the brand names Victoza (diabetes) and Saxenda (weight management). It works by increasing insulin secretion and reducing appetite.

Common Dosages & Protocols:
- Injection: 0.6-3 mg daily subcutaneous injection.

Potential Side Effects:
- Nausea
- Diarrhea
- Pancreatitis (rare)

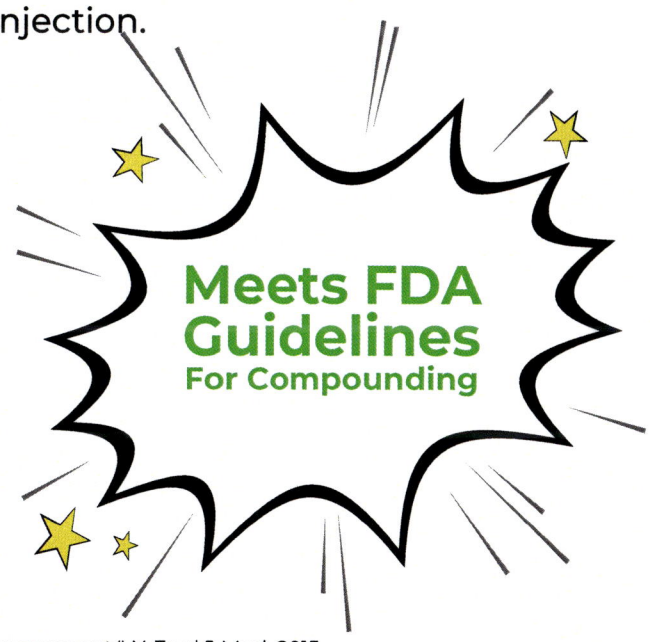

Clinical References:
- Davies MJ, Bergenstal R. "Efficacy of liraglutide for diabetes management." N Engl J Med. 2015.
- Pi-Sunyer FX. "Weight management with liraglutide." Lancet. 2016.

Melanotan II
MT2

Benefits May Include:
- Enhanced skin pigmentation
- Libido enhancement

Melanotan II is a synthetic analog of alpha-melanocyte-stimulating hormone (α-MSH) that stimulates melanin production, resulting in darker skin pigmentation. It has also shown efficacy in enhancing libido and is being researched for potential therapeutic uses.

PT-141 (Bremelanotide) is made from Melanotan II. Bremelanotide is the active metabolite of Melanotan II.

Common Dosages & Protocols:
- Injection: 0.25-1 mg daily subcutaneous injection.
- Nasal Spray: Starting at 1.6 mg in each nostril and up to 10 mg daily

Potential Side Effects:
- Nausea
- Flushing
- Priapism
- Potential increased blood pressure

Clinical References:
- Wessels WH. "Melanotan II in erectile dysfunction research." Int J Impot Res. 2010.
- Hadley ME, Dorr RT. "Melanotan peptides and skin pigmentation." J Invest Dermatol. 2006.

MOTS-c

Mitochondrial Derived Peptide

Benefits May Include:
- Enhanced metabolic function
- Improved energy production
- Insulin sensitivity enhancement
- Increased physical endurance

MMOTS-c is a mitochondria-derived peptide that plays a key role in cellular energy regulation and metabolic health. It influences mitochondrial gene expression, promoting glucose metabolism and protecting against metabolic stress. MOTS-c has been studied for its potential in treating conditions such as obesity, insulin resistance, and age-related metabolic decline.

Common Dosages & Protocols:
- Injection: 5 mg three times a week for 4-6 weeks, then decrease to once weekly for 4 weeks

Potential Side Effects:
- Temporary injection site irritation
- Mild fatigue
- Insomnia
- Increased heart rate

Clinical References:
- Lee C, Zempo H. "MOTS-c: A mitochondrial-encoded peptide regulating metabolic homeostasis." J Mol Endocrinol. 2015.
- Reynolds JC. "Therapeutic potential of MOTS-c in metabolic disorders." Front Endocrinol. 2020.

Oxytocin
Pitocin, Love Hormone

Benefits May Include:
- Enhanced uterine contractions during labor
- Improved social bonding
- Improvement of autism behavior

Oxytocin is a naturally occurring hormone and neurotransmitter that plays a role in social bonding, childbirth, and lactation. It is FDA-approved for inducing labor and controlling postpartum bleeding.

Some studies have found that oxytocin can improve social skills in children with autism, especially those with low baseline levels of the hormone. For example, one study found that children with autism who received oxytocin paid more attention to others' faces during a cooperative game.

Common Dosages & Protocols:
- Injection: Dosages vary based on the use case
- Nasal Spray: 8 to 80 IU; most common is 24 IU daily
- Length varies from 4 consecutive days a week to daily for 24+ weeks

Potential Side Effects:
- Thirst
- Urination
- Constipation
- Runny nose

Clinical References:
- Veening JG, Olivier B. "Oxytocin and social bonding." Front Neuroendocrinol. 2013.
- Tsatsaris V, Carbonne B. "The clinical use of oxytocin." Eur J Obstet Gynecol. 2004.

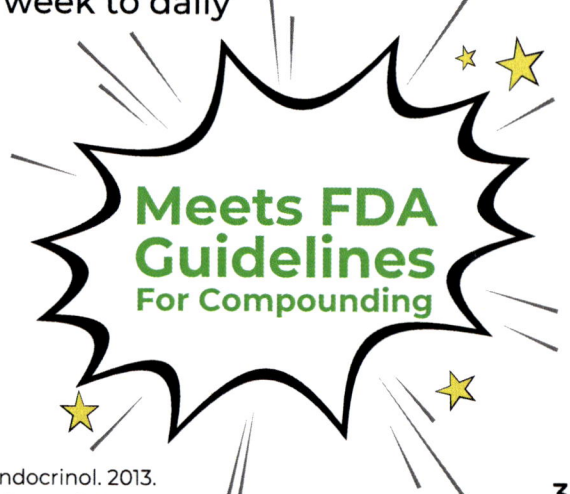

Petrelintide
ZP8396

Benefits May Include:
- Appetite suppression
- Potential weight loss

Petrelintide is a synthetic analog of the hormone amylin, which helps regulate glucose levels and appetite. It is being studied as a potential treatment for obesity due to its effects on appetite suppression and energy expenditure.

Common Dosages & Protocols:
- Injection: Weekly dose 0.6 mg to 9 mg
- Dose increases every 2 weeks

Potential Side Effects:
- Nausea
- Vomiting
- Hypoglycemia (rare)

Clinical References:
- Smith SR, Wadden TA. "The role of amylin analogs in obesity management." Obesity (Silver Spring). 2017.
- Butler PC, Elahi D. "Amylin replacement therapy in diabetes and obesity." Diabetes Obes Metab. 2005.

Retatrutide
Triple GLP-1, GIP, Glucagon Agonist

Benefits May Include:
- Enhanced weight loss
- Improved glucose regulation
- Cardiometabolic health improvement

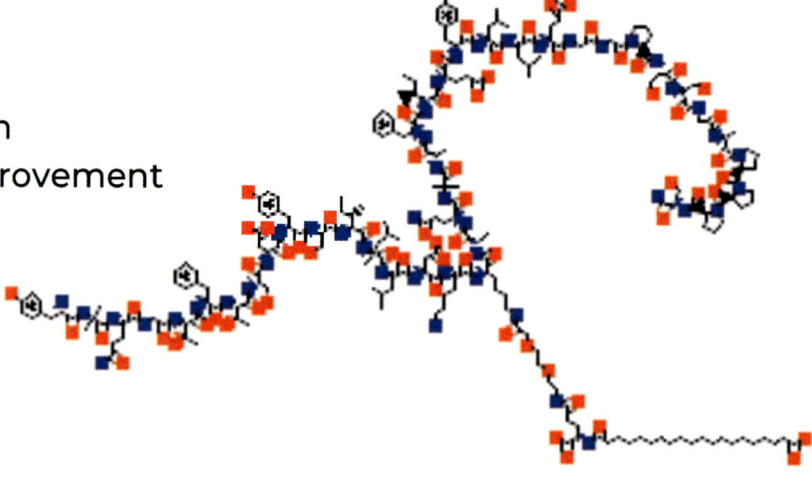

Retatrutide is a novel triple receptor agonist targeting GLP-1, GIP, and glucagon receptors. It is currently under clinical investigation for its potential to promote significant weight loss, improve insulin sensitivity, and enhance cardiometabolic health. Its multi-receptor approach provides synergistic effects, making it a promising candidate in the field of metabolic diseases. Retatrutide has not yet received FDA approval but is progressing through late-stage clinical trials.

Common Dosages & Protocols:
- Injection: Weekly dose 2 mg to 12 mg

Potential Side Effects:
- Nausea
- Diarrhea
- Decreased appetite

Clinical References:
- Jastreboff AM. "Retatrutide: A triple receptor agonist for obesity and metabolic health." Lancet. 2023.
- Drucker DJ. "Advances in GLP-1 therapies and multi-receptor agonists." Nat Rev Drug Discov. 2023.

Selank
TP-7

Benefits May Include:
- Reduced anxiety
- Enhanced cognitive function
- Stress resilience

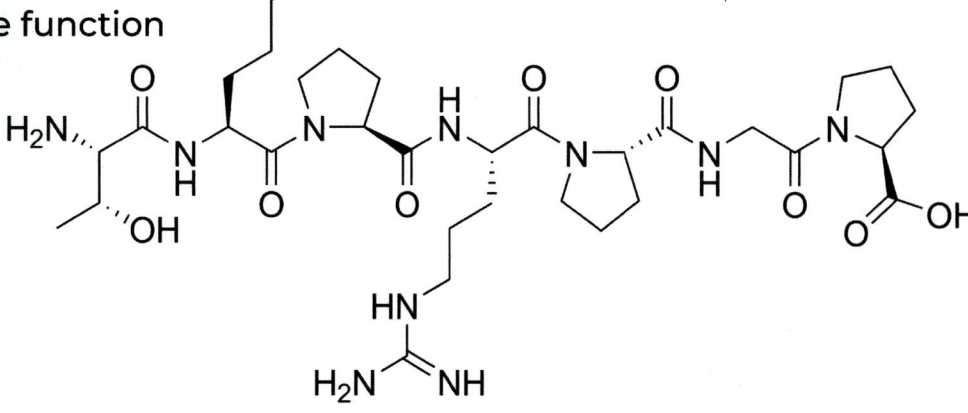

Selank is a synthetic peptide developed for its anxiolytic and nootropic properties. It mimics the effects of naturally occurring immune molecules and has been shown to reduce anxiety while enhancing memory and learning. Selank is not FDA-approved but is widely studied in Eastern Europe.

Common Dosages & Protocols:
- Nasal Spray: 250-500 mcg daily.

Potential Side Effects:
- Mild nasal irritation
- Headaches
- Nausea

Clinical References:
- Andreeva LA, Gurov AN. "Selank in cognitive and anxiety disorders." Neurosci Behav Physiol. 2013.
- Bartis D, Lissin DV. "Peptide-based anxiolytics: Selank and beyond." Curr Med Chem. 2014.

Semaglutide
Ozempic & Wegovy

Benefits May Include:
- Improved blood glucose regulation
- Weight management
- Reduced CV risks

Semaglutide is a GLP-1 receptor agonist approved by the FDA for the treatment of type 2 diabetes (brand name Ozempic) and chronic weight management (brand name Wegovy). It works by mimicking the GLP-1 hormone to regulate blood sugar levels, delay gastric emptying, and suppress appetite.

Common Dosages & Protocols:
- Injection: 0.25-2.4 mg weekly
- Sublingual: 1-2 mg daily
- Nasal: 0.25 mg to 1.5 mg in each nostril daily or 4 x per week

Potential Side Effects:
- Nausea
- Vomiting
- Diarrhea

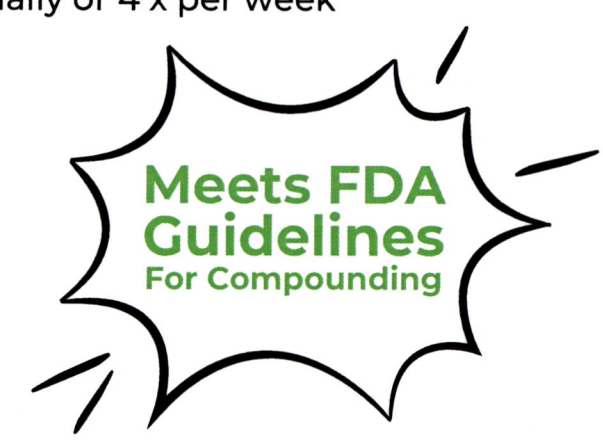

Clinical References:
- Wilding JP. "Semaglutide for weight management and glycemic control." Lancet. 2021.
- Marso SP. "Cardiovascular outcomes with semaglutide in diabetes." N Engl J Med. 2016.

Semax
Semaxum

Benefits May Include:
- Enhanced cognitive performance
- Reduced anxiety

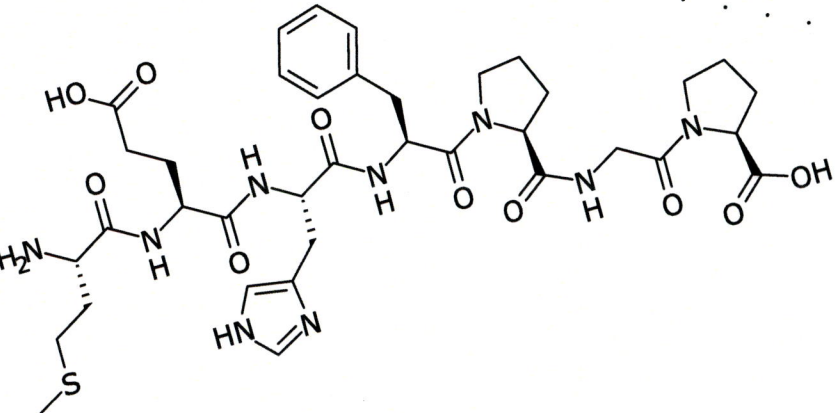

Semax is a synthetic peptide derived from adrenocorticotropic hormone (ACTH). It is widely studied for its nootropic and neuroprotective properties, with applications in cognitive enhancement, stroke recovery, and anxiety management. While not FDA-approved, it is commonly used in Eastern Europe for various neurological conditions.

Semax is categorized as a synthetic nootropic drug and is often compared to more well-known cognitive enhancers like piracetam and modafinil.

Common Dosages & Protocols:
- Nasal Spray: 250-500 mcg daily.

Potential Side Effects:
- Mild nasal irritation

Clinical References:
- Zozulya SA, Sokolova NA. "Semax as a neuroprotective agent." J Neurol Sci. 2011.
- Ashmarin IP. "Nootropic and neuroprotective effects of Semax." Curr Med Chem. 2008..

Sermorelin
Geref

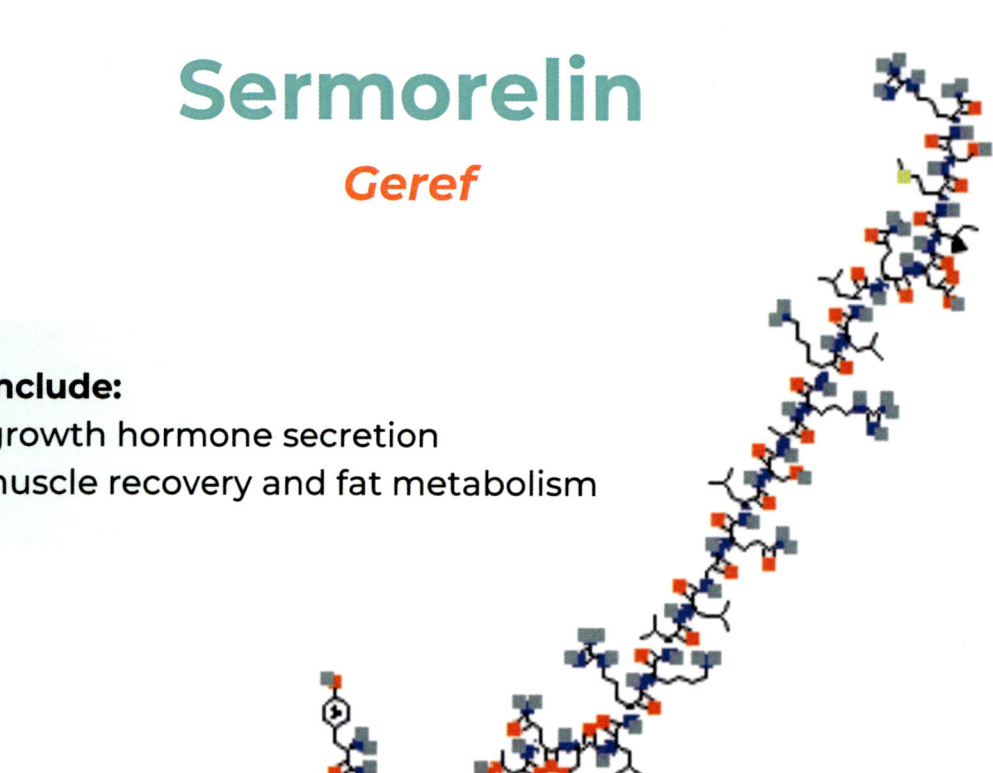

Benefits May Include:
- Enhanced growth hormone secretion
- Improved muscle recovery and fat metabolism

Sermorelin Acetate is a synthetic peptide that stimulates the release of growth hormone by mimicking growth hormone-releasing hormone (GHRH). It is widely used in anti-aging and athletic recovery protocols to promote increased lean muscle mass, energy, and improved recovery times. It was FDA-Approved in 1997 under the brand name Geref. It was discontinued for manufacturing difficulties, not safety or efficacy.

Common Dosages & Protocols:
- Injection: 200-500 mcg daily, subcutaneous.
- Nasal: 10IU to 20IU daily, preferable at night

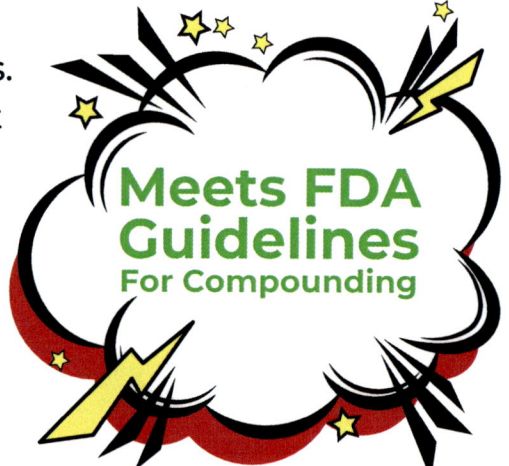

Potential Side Effects:
- Mild headache
- Fatigue
- Joint pain
- Dizziness or nausea
- Brief burning sensation or irritation in the nose

Clinical References:
- Pfaeffle RW. "Therapeutic use of growth hormone-releasing hormone analogs." Horm Res. 2000.
- Wuster C, Senn E. "Sermorelin in growth hormone deficiency management." J Clin Endocrinol Metab. 1995.

TB-500

THYMOSIN Beta-4

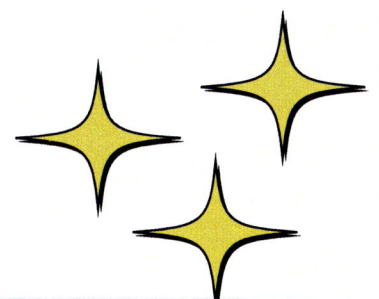

Benefits May Include:
- Accelerated wound healing
- Reduced inflammation
- Enhanced tissue repair

TB-500 is a synthetic version of thymosin beta-4, a naturally occurring peptide that plays a crucial role in tissue regeneration and healing. It is commonly used in research and athletic recovery to promote faster healing of injuries and reduce inflammation. TB-500 has shown promise in preclinical studies but is not FDA-approved.

Common Dosages & Protocols:
- Injection: 5 to 10 mg weekly dose, subcutaneous or intramuscular, depending on protocol. Weekly doses can be split into 2-3 doses per week.

Potential Side Effects:
- Mild redness at the injection site
- Rare allergic reactions
- Nausea
- Vomiting

Clinical References:
- Goldstein AL. "Thymosin beta-4 and tissue repair: Mechanisms and clinical trials." Ann N Y Acad Sci. 2007.
- Malinda KM. "Role of thymosin beta-4 in cellular healing and regeneration." Exp Cell Res. 2010.

Tesamorelin
Egrifta

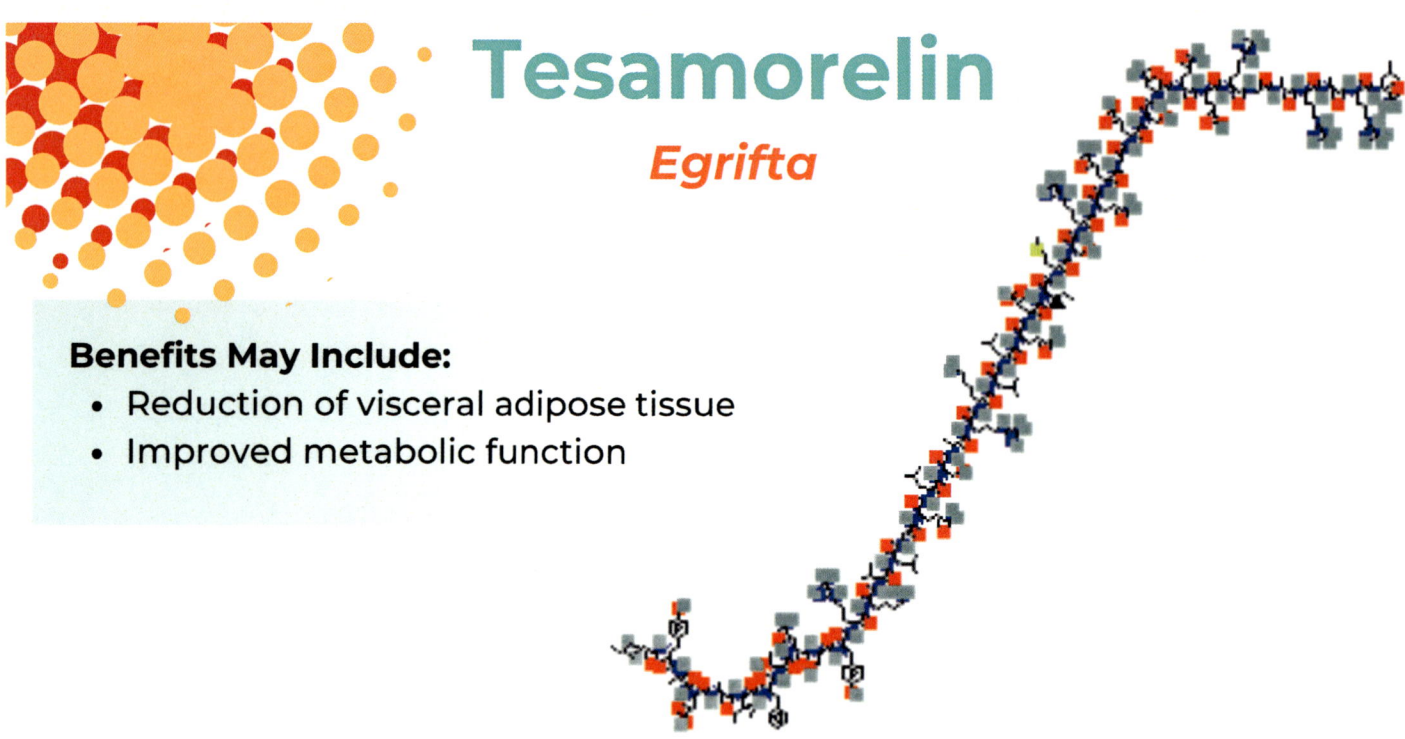

Benefits May Include:
- Reduction of visceral adipose tissue
- Improved metabolic function

Tesamorelin is a synthetic analog of growth hormone-releasing hormone (GHRH) that is FDA-approved under the brand name Egrifta for reducing abdominal fat in HIV-associated lipodystrophy. It works by stimulating the release of growth hormone, which can enhance fat metabolism and improve body composition.

Tesamorelin promotes fat metabolism and helps reduce fat in the abdominal area, leading to a leaner and more defined midsection.

Common Dosages & Protocols:
- Injection: 2 mg subcutaneously daily.

Disqualified From Compounding:
- Tesamorelin meets all the normal rules to compound
- You cannot, due to it being on the biologics list: https://www.fda.gov/media/119229/download

Potential Side Effects:
- Injection site reactions
- Joint pain
- Nausea

Clinical References:
- Falutz J. "Tesamorelin treatment for HIV lipodystrophy." N Engl J Med. 2010.
- Thompson M. "The role of tesamorelin in metabolic health." Curr Opin Endocrinol Diabetes Obes. 2011.

Tesofensine

Serotonin–Noradrenaline–Dopamine Reuptake Inhibitor

Benefits May Include:
- Appetite suppression
- Weight loss

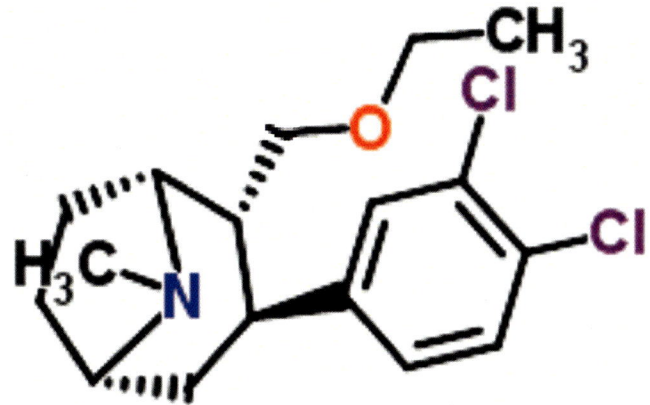

Tesofensine is a triple monoamine reuptake inhibitor (serotonin, norepinephrine, dopamine) initially developed for Parkinson's disease and Alzheimer's disease. It has since shown significant potential as a weight loss agent by reducing appetite and increasing energy expenditure.

Approved with metoprolol as an orphan drug, Tesomet, in 2021. *** *There is some debate on whether the orphan drug status meets the qualifications to compound. Some experts say yes, and some say no.*

Common Dosages & Protocols:
- Oral: 0.25 mg, 0.5 mg, or 1 mg

Potential Side Effects:
- Increased heart rate
- Insomnia
- Dry mouth
- Anxiety
- Increased blood pressure at 1 mg dose

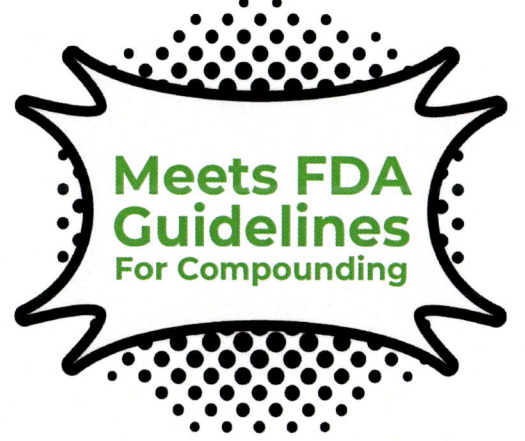

Clinical References:
- Astrup A. "Tesofensine and weight loss: A clinical study." Lancet. 2008.
- Hansen D. "Neurochemical effects of tesofensine." Int J Obes. 2011.

Tirzepatide
GLP-1 and GIP

Benefits May Include:
- Improved blood sugar control
- Significant weight loss

Tirzepatide is a dual GIP and GLP-1 receptor agonist approved by the FDA under the brand name Mounjaro for the treatment of type 2 diabetes. Its unique mechanism promotes enhanced insulin secretion, reduced glucagon levels, and delayed gastric emptying, leading to improved glycemic control and weight management.

Common Dosages & Protocols:
- Injection: 2.5-15 mg weekly.

Potential Side Effects:
- Nausea
- Diarrhea
- Pancreatitis (rare)

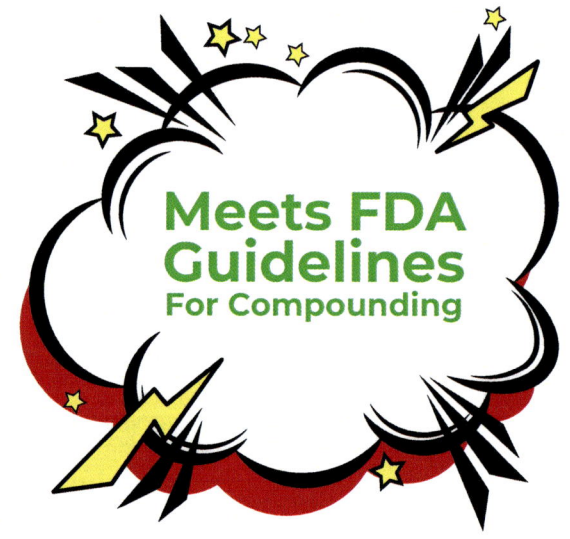

Clinical References:
- Rosenstock J. "Tirzepatide for weight loss and diabetes control." N Engl J Med. 2021.
- Jastreboff AM. "Clinical efficacy of tirzepatide in obesity management." Lancet. 2022.

Zinc Thymulin
Serum Thymic Factor

Benefits May Include:
- Enhanced hair growth
- Immune modulation

Zinc Thymulin is a naturally occurring peptide produced by the thymus gland that requires zinc to be biologically active. It functions as a key regulator of T-cell development and immune function, essentially acting as a hormone that supports the differentiation and activity of different T-cell subsets within the body.

Due to its dependence on zinc, a deficiency in this mineral can significantly impact the effectiveness of thymulin, making it a potential target for treatments related to immune function and, more recently, hair growth. Studies have shown promising results in promoting hair follicle activity and combating hair loss when applied topically.

Common Dosages & Protocols:
- Topical 50mcg/ml apply 1 ml to scalp at bedtime

Potential Side Effects:
- Mild skin irritation at the injection site
- Localized redness or swelling

Clinical References:
- Garcia-Sanchez JE. "The efficacy of Zinc Thymulin in alopecia treatment." Int J Trichology. 2019.
- Dardick I. "Exploring the immunomodulatory effects of Zinc Thymulin." Clin Immunol. 2018.

Popular Peptide Stacks

CJC-1295 + Ipamorelin
Promotes growth hormone release to support muscle growth, fat loss, and recovery.

BPC-157 + TB-500
Enhances tissue repair, reduces inflammation, and accelerates healing processes.

GHRP-6 + CJC-1295
Stimulates growth hormone secretion, aiding in muscle development and fat metabolism.

CJC-1295 + GHRP-2
Boosts growth hormone levels for muscle growth, fat reduction, and vitality.

Selank + Semax
Combines cognitive and neuroprotective benefits to reduce anxiety and enhance focus.

Tesamorelin + Ipamorelin
Supports fat loss and muscle growth with dual growth hormone-releasing effects.

Thymosin Alpha-1 + Thymosin Beta-4
Provides immune modulation alongside tissue repair and anti-inflammatory benefits.

Semaglutide + Tirzepatide
Promotes weight loss and blood sugar regulation for metabolic health.

Kisspeptin-10 + HCG
Supports fertility by enhancing reproductive hormone regulation and ovulation.

AOD9604 + CJC-1295
Boosts fat metabolism and promotes muscle recovery and anti-aging effects.

Popular Peptide Stacks

Ipamorelin + Hexarelin
Provides synergistic stimulation of growth hormone for muscle repair and recovery.

BPC-157 + GHK-Cu
Focuses on enhancing skin healing, reducing inflammation, and promoting collagen production.

Sermorelin + GHRP-6
Stimulates growth hormone release for anti-aging and muscle recovery benefits.

Tesofensine + Tirzepatide/GLP-1
Combines appetite suppression with metabolic health for significant weight loss.

GHRH + GHRP
Combines growth hormone-releasing hormone with growth hormone-releasing peptides to enhance growth hormone secretion.

GHRH + GHRP + IGF-1 LR3
Adds insulin-like growth factor to the GHRH and GHRP stack for muscle growth and recovery.

CJC-1295 + BPC-157 + TB-500
A triple stack for accelerated healing, growth hormone stimulation, and inflammation reduction.

Tesamorelin + GHRP-6
Combines fat loss with appetite regulation and growth hormone release.

CJC-1295 + Melanotan II
Supports anti-aging and enhances skin pigmentation.

Popular Peptide Stacks

AOD9604 + TB-500
Combines fat metabolism with tissue repair and anti-inflammatory properties.

GHK-Cu + IGF-1 LR3
Focuses on skin healing and muscle regeneration.

Thymosin Beta-4 + BPC-157 + IGF-1 LR3
A triple stack for joint healing, tissue repair, and muscle growth.

Semaglutide + Tesofensine
Targets metabolic health with appetite suppression and weight loss.

Melanotan II + BPC-157
Combines skin pigmentation enhancement with wound healing.

GHRP-6 + Hexarelin + CJC-1295
Maximizes growth hormone secretion with complementary peptides.

PT-141 + Kisspeptin-10
Enhances libido and reproductive health.

Sermorelin + Ipamorelin
Dual growth hormone stimulation for anti-aging and recovery.

AOD9604 + GHRP-6 + CJC-1295
A stack for fat metabolism, appetite stimulation, and growth hormone release.

Thymosin Alpha-1 + BPC-157 + TB-500
Combines immune modulation, tissue repair, and inflammation reduction.

OTCs, APIs, & 503Bs...
Additional Resources To Support Your Patients

What options do you have to help your patients when you aren't allowed to compound or dispense most peptides?

What about sourcing APIs for peptides that are ok to compound?

Want to offer sterile options when you are not a sterile compounder?

Below are vetted companies that you can use to help your patients, from raw APIs for compounding to OTC peptide supplements to 503Bs to additional educational resources.

API Sources
- Lone Star Pharmaceutical
- Biopeptek
- Sinopep
- Pharmsource
- Darmerica/Attix

This is not an exhaustive list by any means. I have personally used these companies and would recommend them to a friend.

Research Peptides
- Molecular Solutions Peptides
- Peptide Sciences
- RCS Research

Education
- Science Direct
- Chemistry World
- SSRP Institue
- A4M

OTC Peptides Wholesale
There are many peptides you can't dispense. Buy wholesale and resell OTC.
- **Integrative Peptides** - https://integrativepeptides.com/merchant-registration/
- **InfiniWell** - https://infiniwell.com/pages/register

503Bs
- BPI Labs - john@highhealth.com - 310-779-4996
- PQ Pharmacy - psaid@pqpharmacy.com
- Empower Pharmacy

Bonus!

Comprehensive Alternative Medicine Options Chart

I thought you might enjoy the chart below. It contains peptides and many other products that people use for wellness and optimal health. This chart should help keep you informed on what your patients might be looking up and considering to achieve their health outcomes.

Product Name	Form	Quantity / Concentration	Dosage / Timing	Application Method	Indication
2-Methoxyestradiol	Caps	3mL	100mcg daily before bed	Suppository	Cancer
3-Desoxy DHEA	Caps	100mg (30 Caps)	1 capsule daily	Oral	Aromatase inhibitor
5-amino-1MQ	Inj	10ml at 300mg/ml	150mcg daily for 20 days (30 min post NAD+)	SQ Inj	Weight Loss and Performance
Amlexanox	Caps	40mg (90 Caps)	1 capsule TID	Oral	Insulin sensitivity, weight loss
Ammonium Tetrathiomolybdate	Caps	40mg (90 Caps)	1 capsule TID	Oral	Cancer
Aniracetam	Caps	375mg (60 Caps)	2 Caps daily	Oral	AMPA receptor modulator (Neurogenic)
AOD-9604	Cream	30mL at 600mcg/mL	4 clicks (1mL) daily	Topical	Repair
AOD-9604	Inj	5mL at 1200mcg/mL	0.25mL daily	SQ Inj	Weight loss
Argireline/Leuphasyl	Cream	0.5%/0.5% in 30ml TopiClick	1mL daily	Topical	Reduce wrinkle depth and fine lines
Argireline/GHK-Cu/Leuphasyl	Cream	0.5%/0.2%/0.5% in 30ml TopiClick	1mL daily	Topical	Reduce wrinkle depth and fine lines

Bonus!

Comprehensive Alternative Medicine Options Chart

Product Name	Form	Quantity / Concentration	Dosage / Timing	Application Method	Indication
BPC-157	Inj	3mL at 2000mcg/mL	0.15mL daily	SQ Inj	Repair tendon + ligaments
BPC-157	Caps	500mcg (30 Caps)	1 capsule daily	Oral	Repair bowel/gut
Cerebrolysin	Inj	4x10mL at 215mg/mL	1mL daily	SQ Inj	Neurogenic
CJC 1295	Inj	2M at 2000mcg/mL	0.10mL nightly, 5 nights a week	SQ Inj	GHRH
CJC 1295/Ipamorelin	Inj	2mL (or 5mL) at 1000mcg/mL	0.10mL nightly, 5 nights a week	SQ Inj	GHRH, weight loss
DHH-B	Caps	30 Caps at 7.5mg	1-2 Caps as needed	Oral	Anti-Anxiety
DIHEXA	Caps / Cream	5mg (Caps), 30mL at 20mg/mL	1-2 Caps or 4 clicks (1mL) daily	Oral / Transdermal	Neurological function/repair
DSIP	Inj	3mL at 1000mcg/mL	0.1mL daily, 2-3 times a week	SQ Inj	Endocrine regulation, sleep
Enclomiphene	Caps	25mg (30 Caps)	1 capsule daily	Oral	Aromatase inhibitor (Reproductive system)
Epicatechin	Inj	100mg/mL	100mg daily	SQ Inj	Sarcopenia, Muscle gain

Bonus!

Additional Alternative Medicine Options Chart

Product Name	Form	Quantity / Concentration	Dosage / Timing	Application Method	Indication
Epitalon	Inj	5mL at 3000mcg/mL	0.1mL daily	SQ Inj	Hormone regulation, telomere extension
Fat Loss Cream	Cream	60mL at 0.5% aminophylline/glycrrhetinic acid 2.5%	2 pumps (1mL) twice daily	Transdermal	Fat reduction
FOXO4-DRI	Inj	10ml vial at 10mg/ml	300mcg/kg, 3 doses every other day over 5 days	Intravenous	Clear senescent cells
GHK-Cu	Inj	5mL at 10mg/mL	0.2mL daily	SQ Inj	Skin elasticity
GHK-Cu	Foam	50mL at 5mg/mL	2 pumps (1mL) daily	Topical	Hair loss
IGF-1 LR3	Inj	6.2mL at 100mcg/mL	0.4-0.8mL daily, post workout	SQ Inj	Muscle building, autocrine hormone signaling
Ipamorelin	Inj	5mL at 2000mcg/mL	0.1mL daily, 5 days a week	SQ Inj	GHRH, weight loss
iRGD	Inj	10mL at 2.5mg/mL	40mcg/kg daily, alongside concurrent therapy	SQ Inj	Cancer
Kisspeptin-10	Inj	5mL at 100mcg/mL	0.1mL daily	SQ Inj	LH increase
KPV	Topical	30mL at 15mg/mL	0.5mL applied twice daily	Transdermal cream	Psoriasis

Bonus!

Additional Alternative Medicine Options Chart

Product Name	Form	Quantity / Concentration	Dosage / Timing	Application Method	Indication
Liraglutide	Inj	5mL at 3600mcg/mL	Varies	SQ Inj	Insulin resistance, weight loss
LL-37	Inj	5mL at 2000mcg/mL	Varies with indication	SQ Inj	Antimicrobial
Melanotan 1	Inj	5mL at 2000mcg/mL	0.15mL daily with UVB light exposure	SQ Inj	Vitiligo
Melanotan 2	Inj	5mL at 2000mcg/mL	0.15mL daily for 1-2 weeks, then 0.25mL weekly	SQ Inj	Libido, tanning, weight loss
Met-Enkephalin	Inj	10mg lyophilized vial	10mg IV once weekly	Intravenous	Cancer
MK-677	Caps	25mg (30 Caps)	1 capsule daily, empty stomach	Oral	Growth hormone IGF-1 increase
MOTS-c	Inj	4mL at 10mg/mL	1mL once weekly	SQ Inj	Energy, weight loss
Myristyl	Inj	5mL vial at 6mg/mL	100mcg/kg once weekly	Intravenous	Cholesterol reduction
NMN	Inj	10mL vial at 200mg/mL	200mg (1mL) BID	SQ Inj	Anti-aging, weight loss

Bonus!

Additional Alternative Medicine Options Chart

Product Name	Form	Quantity / Concentration	Dosage / Timing	Application Method	Indication
PEG-MGF	Inj	5mL at 2000mcg/mL	0.1mL daily, 5 days a week	SQ Inj	Muscle building
PNC-27	Inj	5mL at 1000mcg/mL	2.5-5mg daily	SQ Inj	Cancer
Pentosan Polysulfate (PPS)	Inj	10mL at 250mg/mL	250mg	Intramuscular	Osteoarthritis
PT-141	Inj	2mL at 10mg/mL	0.2mL (female), 0.1mL (male) as needed	SQ Inj	Erectile dysfunction, libido
PTD-DBM	Cream	5mL at 0.5% spray	Spray sufficient amount to cover area	Transdermal (after microneedling)	Hair loss
RG3	Nasal	15mL	1 spray per nostril 2-4 times daily	Nasal	Neurogenic
RG3/Methylcobalamin/NAD+	Nasal	15mL	1 spray per nostril 2-4 times daily	Nasal	Neurogenic
SARMS LGD-4033	Caps	10mg (60 Caps)	1 capsule daily	Oral	Muscle building
Selank (Inj)	Inj	5mL at 1000mcg/mL	0.1mL daily	SQ Inj	Neurogenic
Selank (Nasal)	Nasal	3mL at 750mcg/mL	1 spray per nostril daily	Nasal	Neurogenic
Semax	Inj / Nasal	5mL at 1000mcg/mL / 3mL at 7500mcg/mL	0.1mL daily or 1 spray per nostril daily	SQ Inj / Nasal	Neurogenic

Bonus!

Additional Alternative Medicine Options Chart

Product Name	Form	Quantity / Concentration	Dosage / Timing	Application Method	Indication
Tesamorelin	Inj	24 vials (lyophilized) at 1mg/vial	1mg daily, 6 days a week	SQ Inj	Muscle building, weight loss
Tesofensine	Caps	0.5mg (30 Caps)	1 capsule daily	Oral	Weight loss
Thymosin Alpha-1	Inj	3mL at 1000mcg/mL	0.1mL daily	SQ Inj	Immune modulation
Thymosin Beta-4	Inj	3M at 2000mcg/mL	0.1mL daily	SQ Inj	Tissue repair
TMB	Caps	15mg (30 Caps)	1 capsule daily	Oral	Anti-inflammatory
TTA/Amlexanox	Caps	40mg/100mg (90 Caps)	1 capsule TID	Oral	Fat reduction
VIP	Inj	5mL at 100mcg/mL	0.1mL twice daily	SQ Inj	Neurological support
Zinc Thymulin	Foam	30mL at 50mcg/mL	2 pumps (1mL) daily, before bed	Topical	Hair loss

About The Author

Dr. Lisa Faast is a trailblazer in the independent pharmacy industry, with over 20 years of experience as a pharmacy owner, consultant, and entrepreneur. As the CEO of DiversifyRx, she is on a mission to help pharmacy owners love their businesses again by providing proven strategies for growing profits, streamlining operations, and achieving sustainable success.

Known as "The Pharmacy Badass," Dr. Faast combines her no-nonsense approach with actionable advice that has transformed thousands of pharmacies nationwide. She is also the founder of Pharmacy Profit Summit and Pharmacy Badass University, a sought-after speaker, and a dedicated mom of four, inspiring pharmacy owners to thrive both personally and professionally.

In addition to her extensive experience, Dr. Faast is a guest lecturer for the BHRT Academy, where she shares her expertise in bioidentical hormone replacement therapy. She has trained hundreds of healthcare providers on effective weight loss protocols and the clinical use of peptides, cementing her reputation as an authority in this innovative field. Notably, she has provided specialized sterile peptide compounding training to over 50 pharmacy owners, further expanding her impact on the industry.

Dr. Lisa Faast

DiversifyRx
Pharmacy Badass University
Pharmacy Profit Summit
Get Sh*t Done Accelerator

www.drlisafaast.com

Made in United States
Orlando, FL
21 January 2026